YOUR KNOWLEDGE HAS VALUE

Bibliographic information published by the German National Library:

The German National Library lists this publication in the National Bibliography; detailed bibliographic data are available on the Internet at http://dnb.dnb.de .

Imprint:

Copyright © 2017 GRIN Verlag, Open Publishing GmbH
Print and binding: Books on Demand GmbH, Norderstedt Germany
ISBN: 9783668498860

This book at GRIN:

http://www.grin.com/en/e-book/373114/phytochemical-characterization-of-averrhoa-bilimbi-and-in-vitro-analysis

Prem Jose Vazhacharickal, Jiby John Mathew, Sajeshkumar N.K., Nila Joy

Phytochemical characterization of Averrhoa bilimbi and in vitro analysis of cholesterol lowering effect on fatty food materials

GRIN Publishing

GRIN - Your knowledge has value

Since its foundation in 1998, GRIN has specialized in publishing academic texts by students, college teachers and other academics as e-book and printed book. The website www.grin.com is an ideal platform for presenting term papers, final papers, scientific essays, dissertations and specialist books.

Visit us on the internet:

http://www.grin.com/

http://www.facebook.com/grincom

http://www.twitter.com/grin_com

Phytochemical characterization of *Averrhoa bilimbi* and in vitro analysis of cholesterol lowering effect on fatty food materials

Prem Jose Vazhacharickal, Jiby John Mathew, Sajeshkumar N.K and Nila Joy

ACKNOWLEDGEMENTS

Firstly we thank **God Almighty** whose blessing were always with us and helped us to complete this project work successfully.

We wish to thank our beloved Manager **Rev. Fr. Dr. George Njarakunnel,** Respected Principal **Dr. Joseph V.J,** Vice Principal **Fr. Joseph Allencheril,** Bursar **Shaji Augustine** and the Management for providing all the necessary facilities in carrying out the study. We express our sincere thanks to **Mr. Binoy A Mulanthra** (lab in charge, Department of Biotechnology) for the support. This research work will not be possible with the co-operation of many farmers.

Lastly, we extend our indebt thanks to patents, friends, and well wishers for their love and support.

Prem Jose Vazhacharickal*, Jiby John Mathew, Sajeshkumar N.K and Nila Joy

*Address for correspondence
Assistant Professor
Department of Biotechnology
Mar Augusthinose College
Ramapuram-686576
Kerala, India

Table of contents

Table of figures

Table of tables

List of abbreviations

%	: percentage
°C	: Degree celsius
μL	: Microlitre
CCL	: Carbon tetrachloride
DPPH	: Diphenyl1(2, 4, 6 trinitrophenyl) hydrazyl
HDL	: High density lipoprotein
IDL	: Intermediate density lipoprotein
LDL	: Low density lipoprotein
NCEP	: National cholesterol education programme
SE	: Standard error
STZ	: Stretoptozotocin
TC	: Total cholesterol
VLDL	: Very low density lipoprotein

Phytochemical characterization of *Averrhoa bilimbi* and in vitro analysis of cholesterol lowering effect on fatty food materials

Prem Jose Vazhacharickal[1], Jiby John Mathew[1], Sajeshkumar N.K[1] and Nila Joy[1]

[1]Department of Biotechnology, Mar Augusthinose College, Ramapuram, Kerala, India-686576

Abstract

As the prevalence of obesity and hypercholesterolemia are very common in our society, plants with cholesterol lowering action has great value in modern therapeutics. The phytochemicals present in the extracts of *Averrhoa bilimbi* were analyzed and its effect on lowering cholesterol in various fatty food materials was evaluated in vitro. Various phytochemical compounds like tannins, saponins, alkaloids, emodins, proteins, carbohydrate, terpenoids, glycosides, flavonoids, coumarins and phenols were found in the fruit extracts of the plant. The level of cholesterol was evaluated by Zak's method in five different fatty food materials. After the treatment with extract four of them showed significant reduction in the cholesterol level day by day and no change in the cholesterol level was observed in one sample.

Keywords: Cholesterol; Zak's method; Hypercholesterolemia, Antihyperlipidemic, Emodins, Coumarins.

1

1. Introduction

From ancient times plants have provided a source of inspiration for novel drug compounds, plant derived medicines have made large contributions to human health and well-being (Aiyelaagbe et al., 2000). Many of the powerful drugs used in modern medicines originated in plants. Plants are the major source for bio-active compounds they are meant for several biological activities in human and animals having therapeutic properties. Medicinal plants possess various medicinal properties; have been serving as the major sources of therapeutic agents for maintenance of human health (Silver et al., 1990). Besides small molecules from medicinal chemistry, natural products are still major sources of innovative therapeutic agents for various conditions, including infectious diseases (Clardy and Walsh, 2004). Current research on natural molecules and products primarily focuses on plants since they can be sourced more easily and selected on the basis of their ethno-medicinal use (Verpoorte et al., 2005). The lipids in the body are mainly represented by cholesterol, triglycerides, and phospholipids. Elevated blood lipid levels are the major risk factors for the development of cardiovascular diseases, coronary artery disease, cerebro vascular disease, and peripheral vascular disease. These conditions often lead to heart attacks and strokes. Several medicinal plants have been scientifically evaluated for their lipids-lowering property with respect to control aforementioned disorders. Pharmacological investigations have revealed that *Averrhoa bilimbi* possesses lipid-lowering property (Alhassan and Ahmed, 2016).

Averrhoa bilimbi (commonly known as bilimbi, cucumber tree or tree sorrel) is a fruit bearing tree of the genus Averrhoa, family oxalidaceae, is medicinally used as a folk remedy for many symptoms. It is used for the treatment of fever, mumps, pimples, inflammation of the rectum and diabetes, itches, boils, rheumatism, syphilis, bilious colic, whooping cough, hypertension, stomach ache, ulcer and as a cooling drink (Kumar et al., 2011). When used in high concentrations the fruit juice can lead to acute renal failure due to acute tubular-necrosis, owing to its high oxalate contents which results in intra-tubular oxalate crystal deposition.

Cholesterol is carried in the blood attached to proteins called lipoproteins. There are two main forms, LDL (low density lipoprotein) and HDL (high density lipoprotein). LDL cholesterol is often referred to as "bad cholesterol" because too much is unhealthy. HDL is often referred to as "good cholesterol" because it is protective. Now a day many people eating too much saturated fat such as butter, ghee, fatty meat and meat products, full fat cheese, milk, cream, coconut and palm oils and coconut cream that increases cholesterol levels. Many people eating fruits of *Averrhoa bilimbi* and believes that it will reduce the blood cholesterol level.

1.1 Objectives

The objectives of this study to evaluate the phytochemical properties of extract of *Averrhoa bilimbi* fruit pulp and In Vitro analysis of its cholesterol lowering effect on various fatty food materials.

2. Review of literature

From ancient times plants have provided a source of inspiration for novel drug compounds, plant derived medicines have made large contributions to human health and well-being. Many of the powerful drugs used in modern medicines originated in plants. Plants provide wealth of bioactive compounds. They are the main source of drugs that being used from the ancient times as herbal remedies for the health care, prevention and cure of various diseases and ailments (Kalia, 2005). Plants make many chemical compounds for biological functions, including defence against insects, fungi and herbivorous mammals. Over 12,000 active compounds are known to science. These chemicals work on the human body in exactly the same way as pharmaceutical drugs, so herbal medicines can be beneficial and have harmful side effects just like conventional drugs. However, since a single plant may contain many substances, the effects of taking a plant as medicine can be complex. Many Indian plants are used therapeutically for their anti-diabetic effect, anti-hyperlipidemic activity and anti-bacterial activities (Patel et al., 2009). Medicinal plants have been identified and used throughout human history. Plants make many chemical compounds for biological functions, including defence against insects, fungi and herbivorous mammals. At least 12,000 such compounds have been isolated; this is estimated to be less than 10% of the total. These chemicals work on the human body in exactly the same way as pharmaceutical drugs, so herbal medicines can be beneficial and have harmful side effects just like conventional drugs.

Over the past decade, herbal medicine has become a topic of global importance, making an impact on both world health and international trade. Medicinal plants continue to play a central role in the healthcare system of large proportions of the world's population (Akerele, 1988).

Cholesterol is an organic molecule. It is a sterol or modified steroid (Cholesterol, The US National Library of Medicine Medical Subject Headings) a type of lipid molecule, and is biosynthesized by all animal cells, because it is an essential structural component of all animal cell membranes; essential to maintain both membrane structural integrity and fluidity.

Cholesterol is one of three major classes of lipids which all animal cells use to construct their membranes and is thus manufactured by all animal cells. Plant cells do not manufacture cholesterol. It is also the precursor of the steroid hormones and bile acids. Since cholesterol is insoluble in water, it is transported in the blood plasma within protein particles (lipoproteins). Lipoproteins are classified by their density: very low density lipoprotein (VLDL), LDL, intermediate density lipoprotein (IDL) and HDL (Biggerstaff and Wooten, 2004).

Cholesterol enables animal cells to dispense with a cell wall, thereby allowing animal cells to change shape rapidly and animals to move. In addition to its importance for animal cell structure, cholesterol also serves as a precursor for the biosynthesis of steroid hormones, bile acid (Hanukoglu, 1992) and vitamin D. Cholesterol is the principal sterol synthesized by all animals. In vertebrates, hepatic cells typically produce the greatest amounts. It is absent among prokaryotes (bacteria and archaea), although there are some exceptions, such as Mycoplasma, which require cholesterol for growth (Razin and Tully JG, 1970).

Each cell is capable of synthesizing cholesterol by way of a complex 37-step process, beginning with the mevalonate pathway and ending with a 19-step conversion of lanosterol to cholesterol. Furthermore, it can be absorbed directly from animal-based foods.

According to guidelines of national cholesterol education program (NCEP), total cholesterol (TC) concentrations below 200 mg/dL have been regarded as desirable, whereas, concentrations greater than 240 mg/dL are referred to as hyperlipidemic.

Hyper-lipidemia is a condition when abnormally high levels of lipids; the fatty substance are found in the blood. This condition is also called hyper-cholesterolemia/hyper-lipoproteinemia (Amit et al., 2011). The main cause of hyper-lipidemia includes changes in lifestyle habits in which risk factor is mainly poor diet; with a fat intake greater than 40% of total calories, saturated fat intake greater than 10% of total calories; and cholesterol intake greater than 300 mg per day or treatable medical conditions (Durrington, 1995). Hyper-lipidemia associated lipid disorders are considered to cause the atherosclerotic cardiovascular disease. The main aim of treatment in patients with hyper-lipidemia is to reduce the risk of developing ischemic heart disease or the occurrence of further cardio-vascular or cerebro-vascular disease (Smith and Pekkanen, 1992).

Currently used hypo-lipidemic drugs are associated with so many adverse effects and withdrawal is associated with rebound phenomenon which is not seen with herbal preparations. Plant parts or plant extract are sometimes even more potent than known hypo-lipidemic drugs.

Averrhoa bilimbi is a multipurpose, long-lived tropical plant commonly known as"Bilimbi" or "Cucumber Tree" belonging to family Oxalidaceae. The plant has an enormous fiscal value since most of the parts like leaves, bark, flowers, fruits, seeds, roots or the whole plant are used as alternative medicine to treat a variety of diseases.

Averrhoa bilimbi is a native of Moluccas, but the bilimbi is cultivated throughout Indonesia, Philippines, Ceylon and Burma. It is very common in Thailand, Malaya and Singapore; frequent in gardens across the plains of India (Morton, 1987).

Averrhoa bilimbi is a tropical tree, more sensitive to cold especially when very young. It prefers direct sunlight and seasonally humid climates, with evenly distributed rainfall throughout most of the year but there should be a 2-3 month dry season (Roy et al., 2011).

The tree is attractive, long-lived, reaches 16 to 33 feet (5-10 m) in height; has a short trunk soon dividing into a number of upright branches. The wood is white, soft but tough and even-grained

The leaves are mainly clustered at the branch tips. They are alternate, imparipirmate; 30-60 cm long, with alternate or sub opposite leaflets, with rounded base and pointed tip. The leaves are medium-green on the upper surface and pale on the underside. The leaves are applied as a paste on itches, swellings of mumps and rheumatism, and on skin eruption. They are applied on bites of poisonous creatures. Fresh or fermented leaves are used for treatment of venereal disease. A leaf infusion is a remedy for coughs and is taken after childbirth as a tonic. A leaf decoction can be also taken to relieve rectal inflammation

Small, fragrant, 5-petalled flowers, yellowish-green or purplish marked with dark-purple, are borne in small, hairy panicles emerging directly from the trunk and some twigs. A flower infusion is said to be effective against coughs and thrush.

Averrhoa bilimbi is ellipsoid or nearly cylindrical, 4-10 cm long; capped by a thin, star-shaped calyx at the stem-end and tipped with 5 hair-like floral remnants at the apex. The fruit is crisp when unripe, turns from bright-green to yellowish-green, ivory or nearly white when ripe and falls to the ground. The outer skin is glossy, very thin, soft and tender, and the flesh green, jelly-like, juicy and extremely acid. There may be a few flattened, disc-like seeds about 1/4 in (6 mm) wide, smooth and brown. The fruit conserve is administered as a treatment for coughs, beriberi and biliousness. Syrup prepared from the fruit is taken as a cure for fever and inflammation and to stop rectal bleeding and alleviate internal hemorrhoids.

The chemical constituents of *Averrhoa bilimbi* include amino acids, citric acids, cyanidin-3-o-h-glucoside, phenolics, potassium ion and sugars.

Averrhoa bilimbi is a nutrition-packed, starchy fruit that grows mostly on the trunk of tall trees. It is a rich source of vitamin C. Other than the vitamins and minerals, the fruit also consists of fibre, ash, protein and moisture as well as minerals.

Table 1. Nutritional value per 100 g *Averrhoa bilimbi* fruit.

Factor	Value (g)
Moisture	94.2-94.7
Protein	0.61
Fibre	0.6
Ash	0.31-0.40

Table 2. Vitamin content per 100 g *Averrhoa bilimbi* fruit.

Factor	Value (mg)
Vitamin B1	0.010 mg
Riboflavin	0.302
Niacin	0.302
Ascorbic acid	15.6
Carotene	0.035
Vitamin A	0.036

Table 3. Mineral content per 100 g *Averrhoa bilimbi* fruit.

Factor	Value (mg)
Phosphorous	11.1
Calcium	3.4
Iron	1.0

The extract of various parts of *Averrhoa bilimbi* is medicinally used as a folk remedy for many symptoms and showed significant pharmacological activities. As the prevalence of obesity and Diabetes mellitus are very common in our society, research on plants with anti-diabetic and anti-hyperlipidaemic action has great value in modern therapeutics. The data compiled shows that, *Averrhoa bilimbi* is a potent herb for future research since it is anti-hyperlipidaemic. For optimum effect in patients, the components responsible should be isolated, purified and further clinical trials has to be conducted.

Various extracts of fruit and leaves of *Averrhoa bilimbi* have anti-diabetic, anti-microbial, anti-inflammatory, cyto-toxic activities, anti-oxidant activity, anti-fertility, and anti-bacterial activities. These properties of *Averrhoa bilimbi* fruit have been accredited to its saponins, tannins and flavonoids.

The leaves ethanol extract of *Averrhoa bilimbi* was reported to exhibit appreciable antimicrobial activity against six pathogenic microorganisms, namely two Gram positive bacteria (*Bacillus cereus* and *Bacillus megaterium*), two Gram negative bacteria (*Escherichia coli* and *Pseudomonas aeruginosa*), and two fungi (*Aspergillus ochraceous* and *Cryptococcus neoformans*) (Mackeen et al., 1997). The fruit preparations were also found to reduce the microbial load of *Listeria monocytogenes* Scott A and *Salmonella typhimurium* on raw shrimps after washing and during storage (4°C). The fruits and roots extracts of *Averrhoa bilimbi* were also found to exhibit the positive activity against *Mycobacterium tuberculosis* (Mohamad et al., 2011). The leaves extracts have also been reported to display moderate antifungal activity against *Blastomyces dermatitidis, Candida albicans* and *Cryptococcus neoformans,*

Antioxidants are compounds that interact with and neutralize free radicals, thus preventing them from causing cellular damage. It is revealed that *Averrhoa bilimbi* leaves extracts (0.02% w/v) displayed moderate antioxidant activity in ferric thiocyanate and thiobarbituric acid methods while it was found to be inactive in 2, 2 diphenyl1(2, 4, 6 trinitrophenyl) hydrazyl (DPPH) assay. Unlike the leaves, the fruits extracts showed a strong DPPH radical scavenging activity (Achat et al., 2012).

The liver is a vital organ in the body which performs a major role in the metabolism, secretion, storage, and detoxification of chemical substances. Hepato-protective activity is the ability of a compound or extract to prevent liver damage. The methanol extract of *Averrhoa bilimbi* leaves exhibited appreciable hepato-protective activity against carbon tetrachloride (CCl) induced liver toxicity in Wistar rats (Thamizh Selvam et al., 2015).

Cytotoxicity assays are used to determine whether a compound or extract is toxic to cells. Ethanol extract of *Averrhoa bilimbi* leaves has been shown to possess moderate cytotoxic activity (LC, 5.81 µg/l) in brine shrimp lethality assay (Karon et al., 2011).

Several medicinal plants have been shown to possess a significant healing effect. In this regard, the use of *Averrhoa bilimbi* in treating oral injuries has been scientifically investigated as well. The result obtained showed that application of ethanol extract of *Averrhoa bilimbi* leaves (10% concentration) enhanced gingival wound healing (Nair et al., 2014).

Scientific investigations revealed that *Averrhoa bilimbi* possesses anti-diabetic properties. Pushparaj et al. (2000) evaluated the hypo-glycemic and hypo-lipidemic effects of *Averrhoa bilimbi* leaves extract in streptozotocin (STZ) induced diabetic rats. In this study, it was observed that the leaves ethanol extract (125 mg/kg twice daily p.o.) significantly lowered blood glucose and triglyceride levels when compared with the vehicle (Pushparaj et al., 2000).

Elevated blood lipid levels are the major risk factors for the development of cardiovascular diseases, coronary artery disease, cerebro-vascular disease, and peripheral vascular disease. These conditions often lead heart attacks and strokes. Pharmacological investigations have revealed that *Averrhoa bilimbi* possesses lipid lowering property (Pattamadilok et al., 2015).

Hypertension is considered a major risk factor for several cardiovascular diseases such as atherosclerosis, heart failure, stroke, coronary artery disease, and renal insufficiency. Recently, attention has been greatly concentrated on the use of herbal preparations as alternative agents to cure and prevent cardiovascular complications. Traditionally, the fruits and leaves of *Averrhoa bilimbi* have also been efficaciously used for blood pressure symptom. It was observed that the leaves aqueous extract significantly decreased the contractility of the norepinephrine stimulated guinea pig atria without affecting their beating frequency (Bipat et al., 2008). The leaves extract also demonstrated a significant antihypertensive effect in an in vivo experiment using cats revealing the ability of leaves extract to become a potential antihypertensive drug.

Anticoagulant herbs are used as an antithrombotic agent. Anticoagulant herbs are efficaciously used in angina, hepatitis, coronary artery disease, dysmenorrhea, rheumatoid arthritis, traumatic injury, tumours, depression, renal failure, stroke prevention, and post stroke syndrome. The anticoagulant activity of *Averrhoa bilimbi* was reported in normal and alloxan induced diabetic rats (Daud et al., 2013).

3. Hypothesis
The current research work is based on the following hypothesis
1) *Averrhoa bilimbi* extracts are rich in various phytochemical components.
2) These extract could lower cholesterol levels.

4. Materials and Methods

4.1 Study area
Kerala state covers an area of 38,863 km² with a population density of 859 per km² and spread across 14 districts. The climate is characterized by tropical wet and dry with average annual rainfall amounts to 2,817 ± 406 mm and mean annual temperature is 26.8°C (averages from 1871-2005; Krishnakumar et al., 2009). Maximum rainfall occurs from June to September mainly due to South West Monsoon and temperatures are highest in May and November.

4.2 Collection of plant material

The fruits of *Averrhoa bilimbi* were collected from Ramapuram village, Meenachil Taluk, Kottayam district, Kerala (Survey no.183/4,). It was identified taxonomically and stored.

4.3 Preparation of *Averrhoa bilimbi* fruit pulp extracts

The fruits of the plants were cleaned properly. A half of the collected fruit was dried and subjected to Soxhlet extraction and the other half was homogenized in a blender. Both the extract was filtered using Whatman filter paper No.42. The clear filtrate was obtained and was stored in the refrigerator at 4°C for experimental purpose.

4.4 Phytochemical screening

Chemical tests were carried out on the aqueous extract using standard procedures to identify the constituents as described by Sofowara (1993), Trease and Evans (1989) and Harborne (1973).

4.4.1 Test for alkaloids

Two ml of plant extract was taken in a test tube and few drops of Hager's reagent were added. Yellow precipitate shows positive result for alkaloids.

4.4.2 Test for anthraquinones

Three ml of plant extract was taken in a test tube and three ml of benzene and five ml of ten percentage NH_3 were added. Formation of pink, violet or red coloration in ammonical layer detect the presence of anthraquinones.

4.4.3 Test for anthocyanins

Two ml of plant extract was taken in a test tube and two ml of 2N HCl and NH_3 were added. Formation of pinkish red to bluish violet coloration indicates the presence of anthocyanins.

4.4.4 Test for carbohydrate

Two ml of plant extract was taken in a test tube and ten ml of water, two drops of twenty percentage ethanolic α naphthol and two ml of conc.H_2SO_4 were added. Formation of reddish violet ring at the junction shows the presence of carbohydrates.

4.4.5 Test for coumarins

Two ml of extract was taken in a test tube and three ml of ten percentage NaOH was added. Formation of yellow colour gives positive result to coumarins.

4.4.6 Test for emodins

Two ml of plant extract was taken in a test tube and two ml of NH_4OH and three ml of benzene were added. Formation of red colour indicates the presence of emodins.

4.4.7 Test for flavonoids

Five ml of dilute ammonia solution were added to a portion of the plant extract followed by addition of concentrated H_2SO_4. A yellow colouration observed in each extract indicated the presence of flavonoids. The yellow colouration disappeared on standing.

4.4.8. Test for glycosides

Two ml of plant extract was taken in a test tube and two ml of chloroform and two ml of acetic acid were added. Formation of violet to blue to green coloration shows the presence of glycosides.

4.4.9 Test for leucoanthocyanins

Five ml of isoamyl alcohol taken in a test tube and five ml of plant extract was added. Turn organic layer into red detects the presence of leucoanthocyanins.

4.4.10 Test for phlobatannins

Deposition of a red precipitate when an extract of each plant sample was boiled with one percentage aqueous hydrochloric acid was taken as evidence for the presence of phlobatannins.

4.4.11 Test for proteins

One ml of plant extract was mixed with one ml of conc.H_2SO_4 in a test tube. Formation of white precipitate indicate the presence of proteins.

4.4.12 Test for phenols

Few ml of the plant extract was taken in attest tube and few ml of lead acetate was added to it. Formation of white precipitate detects the presence of phenols.

4.4.13 Test for saponins

Ten ml of the extract was mixed with five ml of distilled water and shaken vigorously for a stable persistent froth. The frothing was mixed with three drops of olive oil and shaken vigorously, then observed for the formation of emulsion.

4.4.14 Test for steroids

Two ml of extract was taken in a test tube and two ml chloroform and two ml of conc.H_2SO_4 was added. Formation of reddish brown ring at the junction shows the presence of steroids.

4.4.15 Test for terpenoids

Five ml of each extract was mixed in two ml of chloroform, and concentrated H_2SO_4 (three ml) was carefully added to form a layer. A reddish brown colouration of the inter face was formed to show positive results for the presence of terpenoids.

4.5 Preparation of cholesterol samples

One gram of the sample was dissolved in one ml of chloroform and stored in brown bottle for further use (Varley, 2004).

4.6 Treatment

200 µl of extract was added to each of the sample prepared and mixed well. These are used for the periodic (24 hour interval) determination of cholesterol.

4.7 Estimation of cholesterol

The amount of cholesterol in each sample was estimated by Zak's method before and after treatment (Varley, 2004).

4.8 Statistical analysis

The survey results were analyzed and descriptive statistics were done using SPSS 12.0 (SPSS Inc., an IBM Company, Chicago, USA) and graphs were generated using Sigma Plot 7 (Systat Software Inc., Chicago, USA).

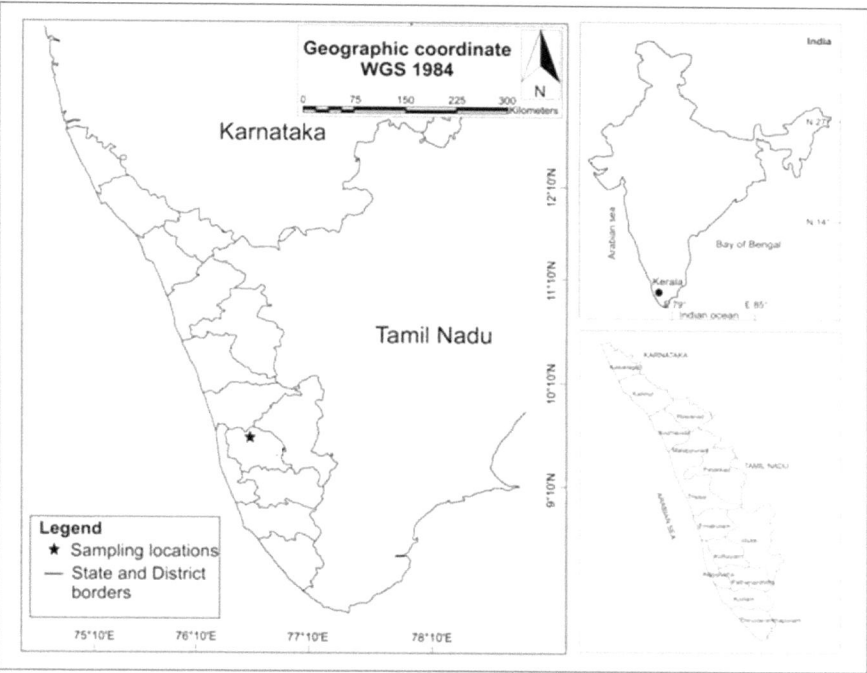

Figure 1. Map of Kerala showing the soil sample collection point. Authors own work.

13

Figure 2. Details of *Averrhoa bilimbi* plant with fruits a), e), f) fruits on tree trunk, b) mature fruits, c), d) fruits cut opened, g) fruit cross section. Authors own image.

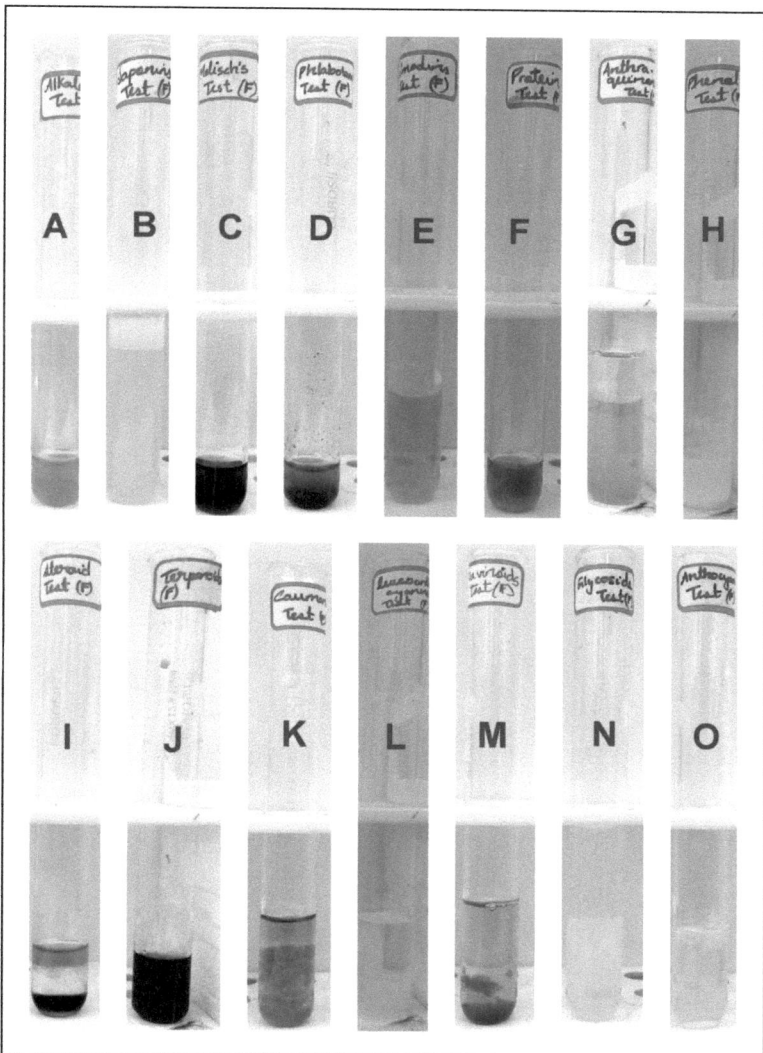

Figure 3. Details of phytochemical analysis of *Averrhoa bilimbi* fruit (water extract S1); A. terpenoids, B. leucoanthocyanins, C. flavonoids, D. proteins, E. alkaloids, F. anthraquinones, G. glycosides, H. coumarins, I. emodin J. carbohydrate, K. phenols, L. saponins, M. phlobatannins, N. steroids, O. anthocyanins. Authors own image.

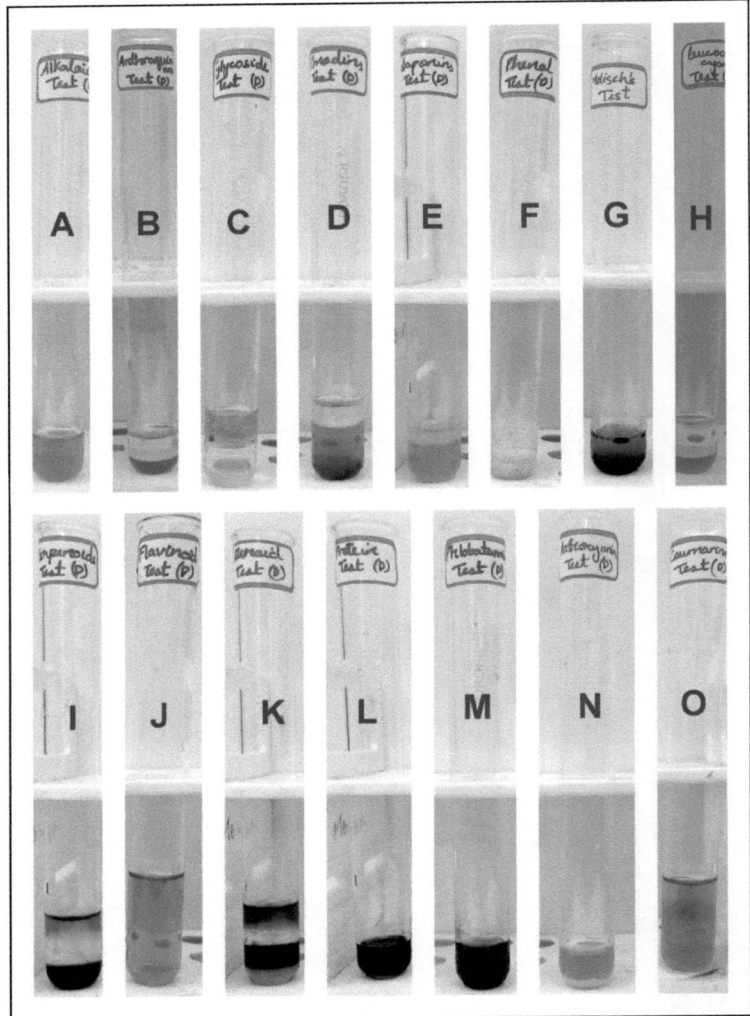

Figure 4. Details of phytochemical analysis of *Averrhoa bilimbi* fruit (water extract S2); A. terpenoids, B. leucoanthocyanins, C. flavonoids, D. proteins, E. alkaloids, F. anthraquinones, G. glycosides, H. coumarins, I. emodin J. carbohydrate, K. phenols, L. saponins, M. phlobatannins, N. steroids,   O. anthocyanins. Authors own image.

Table 4. Preliminary phytochemical analysis of *Averrhoa bilimbi* fruit extracts.

Sl.No	Phytoconstituents	*Averrhoa bilimbi* (Dried)	*Averrhoa bilimbi* (Fresh)
1.	Alkaloid	+	+
2.	Anthocyanins	-	-
3.	Anthraquionones	-	-
4.	Carbohydrates	+	+
5.	Coumarins	+	+
6.	Emodins	-	-
7.	Flavonoids	+	+
8.	Glycoside	-	-
9.	Leucoanthocyanins	-	-
10.	Phenols	+	+
11.	Phlobatannins	+	+
12.	Protein	+	+
13.	Saponins	+	+
14.	Steroids	+	+
15.	Terpenoids	+	+

+ indicates presence of phytochemicals
- indicates absence of phytochemicals

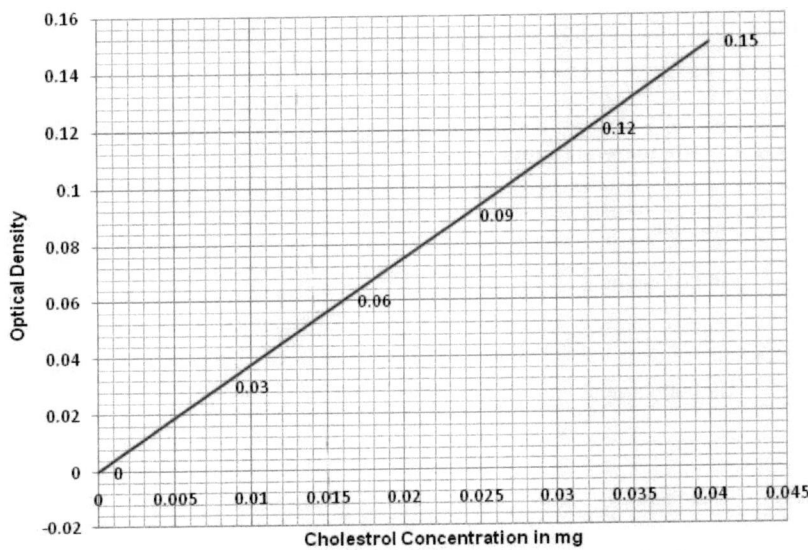

Figure 5. Standard graph for cholesterol estimation by Zak's method.

Phytochemical characterization of *Averrhoa bilimbi* and in vitro analysis of cholesterol lowering effect on fatty food materials

Table 5. Cholesterol estimation at different time intervals after treatment (n=3; values in mg/g sample).

Day	Substrate				
	Egg yolk	Pork Fat	Chicken Fat	Ghee	Cod liver oil
1	0.024 ± 0.01	0.024 ± 0.01	0.026 ± 0.03	0.024 ± 0.02	0.013 ± 0.03
2	0.024 ± 0.01	0.016 ± 0.01	0.026 ± 0.02	0.021 ± 0.01	0.013 ± 0.01
3	0.024 ± 0.01	0.013 ± 0.02	0.024 ± 0.01	0.021 ± 0.01	0.012 ± 0.01
4	0.024 ± 0.01	0.013 ± 0.01	0.018 ± 0.01	0.016 ± 0.01	0.012 ± 0.01
5	0.021 ± 0.01	0.013 ± 0.00	0.016 ± 0.01	0.016 ± 0.01	0.012 ± 0.00
6	0.018 ± 0.02	0.010 ± 0.01	0.016 ± 0.01	0.013 ± 0.01	0.012 ± 0.00
7	0.018 ± 0.01	0.010 ± 0.00	0.013 ± 0.01	0.013 ± 0.01	0.012 ± 0.00
8	0.018 ± 0.02	0.008 ± 0.01	0.013 ± 0.01	0.013 ± 0.01	0.012 ± 0.01
9	0.018 ± 0.00	0.008 ± 0.01	0.013 ± 0.00	0.013 ± 0.01	0.012 ± 0.01
10	0.018 ± 0.01	0.008 ± 0.01	0.013 ± 0.01	0.013 ± 0.01	0.012 ± 0.01
11	0.010 ± 0.01	0.008 ± 0.00	0.013 ± 0.00	0.010 ± 0.01	0.012 ± 0.01
12	0.008 ± 0.01	0.008 ± 0.00	0.013 ± 0.00	0.010 ± 0.00	0.012 ± 0.00
13	0.005 ± 0.01	0.005 ± 0.00	0.005 ± 0.00	0.008 ± 0.01	0.012 ± 0.01
14	0.002 ± 0.01	0.002 ± 0.00	0.005 ± 0.00	0.008 ± 0.00	0.012 ± 0.00

Numbers represent means ± one standard error (SE) of mean.

5. Results and discussion

The phytochemical screening of the extract of *Averrhoa bilimbi* revealed the presence of compounds like alkaloids, coumarins, carbohydrates, flavonoids, proteins, phenol, phlobatannins, saponins, steroids and terpenoids.

The presence of flavonoids and phenols indicated that *Averrhoa bilimbi*, has potential antioxidant activity due to the synergistic effect of these phytoconstituents (Thamizh Selvam et al., 2015). The present study has well demonstrated the phytochemical diversity of *Averrhoa bilimbi* fruit extract. The presence of flavonoids, coumarins, saponins and phenols was well established in the present study which is also at par with available reports from different regions (Kumar et al., 2013). The presence of primary metabolites like carbohydrate and protein and secondary metabolites like flavanoid, alkaloid, phenol and coumarin was reported from *Averrhoa bilimbi* leaves (Alisha and Raphael, 2016). Similar observations are reported by Yan et al. (2013) too. In short, *Averrhoa bilimbi* fruit can be recommended as a neutraceutical functional food.

The in vitro analysis shows that cholesterol level of fatty food materials treated with the extract show reducing day by day except cod liver oil. There was no change in the cod liver oil. The early days there is no notable change observed but after one week of incubation the cholesterol level tremendously lowered.

It has suggested that phytosterols are responsible for phytochemical-related lipid reduction. Phytosterols are a subclass of phytochemicals with potent lipid lowering properties; several studies have evaluated cholesterol-lowering effect of phytosterols (Rayalam et al., 2008; Lerman et al., 2010). Studies showed that enrichment of food products with phytosterols could effectively improve lipid and lipoprotein levels (Katan et al., 2003; Demonty et al., 2008).

Averrhoa bilimbi is medicinally important and is used as a folk remedy for many symptoms and showed significant pharmacological activities. The presence of immense phytochemicals in the plant can be utilized for production of novel drugs to combat various diseases.

Further studies are required to provide the exact mechanism of action of the reported activities by using purified fraction of the same. The further research is very much required in the aspects of isolation and characterization of potential compounds from the extract of *Averrhoa bilimbi* and validating them for the development of new drugs.

6. Conclusions

The study was conducted to analyses the phytochemical characteristics of *Averrhoa bilimbi* and evaluates its cholesterol lowering activity. *Averrhoa bilimbi* is medicinally used as a folk remedy for various symptoms. Most of the parts like leaves, bark, flowers, fruits, seeds, roots or the whole plant are used as alternative medicine to treat a variety of diseases.

Phytochemical tests were carried out on the aqueous extract and on the powdered specimens using standard procedures to identify the constituents. Various phytochemical compounds like tannins, saponins, alkaloids, emodins, proteins, carbohydrate, terpenoids, glycosides, flavonoids, coumarins and phenols were found in the fruit extracts of the plant.

Five common fatty food materials like egg yolk, pork fat, chicken fat, ghee and cod liver oil was mixed with the fruit extract of *Averrhoa bilimbi* and the cholesterol level in each was evaluated in vitro, after and before treatment. The in vitro analysis shows that cholesterol level of fatty food materials treated with the extract show reducing day by day except cod liver oil. The change in cholesterol level was not evident in the early days but significant reduction was observed after a week.

The results show that, *Averrhoa bilimbi* is a potent herb for future research since it has anti-hyperlipidemic properties. For optimum effect in patients, the components responsible should be isolated, purified and further clinical trials has to be conducted. Hence, further studies are recommended to be undertaken to isolate the exact compound(s) and to better recognize the mechanism of such actions scientifically.

Acknowledgements

The authors are grateful for the cooperation of the management of Mar Augusthinose college for necessary support. Technical assistance from Binoy A Mulanthra is also acknowledged.

References

Achat, S., Rakotomanomana, N., Madani, K., & Dangles, O. (2016). Antioxidant activity of olive phenols and other dietary phenols in model gastric conditions: Scavenging of the free radical DPPH and inhibition of the haem-induced peroxidation of linoleic acid. *Food Chemistry, 213*, 135-142.

Aiyelaagbe, O. O., Adesogan, E. K., Ekundayo, O., & Adeniyi, B. A. (2000). The antimicrobial activity of roots of Jatropha podagrica (Hook).*Phytotherapy Research, 14*(1), 60-62.

Akerele, O. (1988). Medicinal plants and primary health care: an agenda for action. *Fitoterapia, 59*(5), 355-363.

Alhassan, A. M., & Ahmed, Q. U. (2016). Averrhoa bilimbi Linn.: A review of its ethnomedicinal uses, phytochemistry, and pharmacology. *Journal of pharmacy & bioallied sciences, 8*(4), 265.

Asna, A. N., & Noriham, A. (2014). Antioxidant activity and bioactive components of oxalidaceae fruit extracts. *Malaysian Journal of Analytical Sciences, 18*(1), 116-126.

Biggerstaff, K. D., & Wooten, J. S. (2004). Understanding lipoproteins as transporters of cholesterol and other lipids. *Advances in Physiology Education, 28*(3), 105-106.

Bipat, R., Toelsie, J. R., Joemmanbaks, R. F., Gummels, J. M., Klaverweide, J., Jhanjan, N., & Ramjiawan, K. (2008). Effects of plants popularly used against hypertension on norepinephrine-stimulated guinea pig atria. *Phcog Mag, 4*(13), 12-19.

Bipat, R., Toelsie, J. R., Joemmanbaks, R. F., Gummels, J. M., Klaverweide, J., Jhanjan, N., & Ramjiawan, K. (2008). PHCOG MAG.: Research Article Effects of plants popularly used against hypertension on norepinephrine-stimulated guinea pig atria. *Phcog Mag, 4*(13), 12.

Chowdhury, S. S., Uddin, G. M., Mumtahana, N., Hossain, M., & Hasan, S. R. (2012). In-vitro antioxidant and cytotoxic potential of hydromethanolic extract of Averrhoa bilimbi L. fruits. *International Journal of Pharmaceutical Sciences and Research, 3*(7), 2263-2268.

Clardy, J., & Walsh, C. (2004). Lessons from natural molecules. *Nature,432*(7019), 829-837.

Daud, N., Hashim, H., & Samsulrizal, N. (2013). Anticoagulant Activity of Averrhoa Bilimbi Linn in Normal and Alloxan-Induced Diabetic Rats. In *Open Conf Proc J* (Vol. 4, No. Suppl 2, M6, pp. 21-6).

Daud, N., Hashim, H., & Samsulrizal, N. (2013). Anticoagulant Activity of Averrhoa Bilimbi Linn in Normal and Alloxan-Induced Diabetic Rats. In *Open Conf Proc J* (Vol. 4, No. Suppl 2, M6, pp. 21-6).

Demonty, I., Ras, R. T., van der Knaap, H. C., Duchateau, G. S., Meijer, L., Zock, P. L., & Trautwein, E. A. (2008). Continuous dose-response relationship of the LDL-cholesterol–lowering effect of phytosterol intake. *The Journal of Nutrition*, 139(2):271–84.

Durrington, P. N. (1995). Lipoprotein (a). *Baillière's clinical endocrinology and metabolism*, 9(4), 773-795.

Hanukoglu, I. (1992). Steroidogenic enzymes: structure, function, and role in regulation of steroid hormone biosynthesis. *The Journal of steroid biochemistry and molecular biology*, 43(8), 779-804.

Harborne, J. B. (1973). Phenolic compounds. In *Phytochemical methods* (pp. 33-88). Springer Netherlands.

Hartini, I. G. A. A. (2012). Topical application of ethanol extract of starfruit leaves (Averrhoa bilimbi linn) increases fibroblasts in gingival wounds healing of white male rats. *Indonesian Journal of Biomedical Sciences*, 6(1), 35-39.

Kalia, A. N. (2005). Text Book of Industrial Pharmacognosym. Oscar Publication, New Delhi,India

Karon, B., Ibrahim, M., Mahmood, A., Huq, A. K. M. M., Chowdhury, M. M. U., Hossain, M. A. H., & Rashid, M. A. (2011). Preliminary antimicrobial, cytotoxic and chemical investigations of Averrhoa bilimbi Linn. and Zizyphus mauritiana Lam. *Bangladesh Pharmaceutical Journal*, 14(2), 127-131.

Katan, M. B., Grundy, S. M., Jones, P., Law, M., Miettinen, T., Paoletti, R., & Participants, S. W. (2003). Efficacy and safety of plant stanols and sterols in the management of blood cholesterol levels. In *Mayo Clinic Proceedings* (Vol. 78, No. 8, pp. 965-978). Elsevier.

Krishnakumar, K. N., Rao, G. P., & Gopakumar, C. S. (2009). Rainfall trends in twentieth century over Kerala, India. *Atmospheric Environment, 43*(11), 1940-1944.

Kumar, A. S., Kavimani, S., & Jayaveera, K. N. (2011). A review on medicinal plants with potential antidiabetic activity. *International journal of phytopharmacology, 2*(2), 53-60.

Lai, P. K., & Roy, J. (2004). Antimicrobial and chemopreventive properties of herbs and spices. *Current medicinal chemistry, 11*(11), 1451-1460.

Lerman, R. H., Minich, D. M., Darland, G., Lamb, J. J., Chang, J. L., Hsi, A., ... & Tripp, M. L. (2010). Subjects with elevated LDL cholesterol and metabolic syndrome benefit from supplementation with soy protein, phytosterols, hops rho iso-alpha acids, and Acacia nilotica proanthocyanidins. *Journal of Clinical Lipidology, 4*(1), 59-68.

Mackeen, M. M., Ali, A. M., El-Sharkawy, S. H., Manap, M. Y., Salleh, K. M., Lajis, N. H., & Kawazu, K. (1997). Antimicrobial and cytotoxic properties of some Malaysian traditional vegetables (ulam). *International Journal of Pharmacognosy, 35*(3), 174-178.

Mohamad, S., Zin, N. M., Wahab, H. A., Ibrahim, P., Sulaiman, S. F., Zahariluddin, A. S. M., & Noor, S. S. M. (2011). Antituberculosis potential of some ethnobotanically selected Malaysian plants. *Journal of Ethnopharmacology, 133*(3), 1021-1026.

Morton, J. 1987. Bilimbi. p. 128–129 In: Fruits of warm climates. Julia F. Morton, Miami, FL.

Morton, J. 1987. Bilimbi. p. 128–129 In: Fruits of warm climates. Julia F.Morton, Miami, FL

Nair, S., George, J., Kumar, S., & Gracious, N. (2014). Acute oxalate nephropathy following ingestion of Averrhoa bilimbi juice. *Case reports in nephrology, 2014.*

Patel, S. S., Shah, R. S., & Goyal, R. K. (2009). Antihyperglycemic, antihyperlipidemic and antioxidant effects of Dihar, a polyherbal ayurvedic formulation in streptozotocin induced diabetic rats. Indian Journal of Experimental Biology, 47(1), 564- 570.

Pattamadilok, D., Niumsakul, S., Limpeanchob, N., Ingkaninan, K., & Wongsinkongman, P. (2010). Screening of cholesterol uptake inhibitor from Thai medicinal plant extracts. *Journal of Traditional Thai and Alternative Medicine, 8*(2/3), 146-151.

Pushparaj, P., Tan, C. H., & Tan, B. K. H. (2000). Effects of *Averrhoa bilimbi* leaf extract on blood glucose and lipids in streptozotocin-diabetic rats. *Journal of Ethnopharmacology, 72*(1), 69-76.

Rayalam, S., Della-Fera, M. A., & Baile, C. A. (2008). Phytochemicals and regulation of the adipocyte life cycle. *The Journal of Nutritional Biochemistry,19*(11), 717-726.

Roy, A., Geetha, R. V., & Lakshmi, T. (2011). Averrhoa bilimbi Linn'Nature's Drug Store-A Pharmacological Review. *International Journal of Drug Development and Research*, 3(3), 101-106.

Smith, G. D., & Pekkanen, J. (1992). Should there be a moratorium on the use of cholesterol lowering drugs?. *BMJ: British Medical Journal, 304*(6824), 431.

Sofowora, A. (1993). Recent trends in research into African medicinal plants.*Journal of Ethnopharmacology, 38*(2-3), 197-208.

Tapsell, L. C., Hemphill, I., Cobiac, L., Sullivan, D. R., Fenech, M., Patch, C. S., & Fazio, V. A. (2006). Health benefits of herbs and spices: The past, the present, the future. *Medical Journal of Australia,185*(4), S1.

Thamizh Selvam, N., Santhi, P. S., Sanjayakumar, Y. R., Venugopalan, T. N., Vasanthakumar, K. G., & Swamy, G. K. (2015). Hepatoprotective activity of Averrhoa bilimbi fruit in acetaminophen induced hepatotoxicity in Wistar albino rats. *Journal of Chemical and Pharmaceutical Research*, 7(1), 535-540.

Thamizhselvam, N., Liji, I. V., Sanjayakumar, Y. R., Sanal Gopi, C. G., & Vasantha Kumar, K. G. (2015). Evaluation of Antioxidant Activity of Averrhoa bilimbi Linn. *Fruit Juice in Paracetamol Intoxicated Wistar Albino Rats. Enliven: Toxicology & Allied Clinical Pharmacology*, 1(1), 002.

Trease, G. E., & Evans, W. C. Pharmacognosy. 1989. *Bailliere Tindall, London*, 45-50.

Varley, H. (2004). Practical clinical Biochemistry, 4th edition, Heinemann Medical, UK.

Verpoorte, R., Choi, Y. H., & Kim, H. K. (2005). Ethnopharmacology and systems biology: a perfect holistic match. *Journal of Ethnopharmacology, 100*(1), 53-56.